Abdelhafid Mimouni

Wernicke-Korsakoff: Bioinorganic Vision

Abdelhafid Mimouni

Wernicke-Korsakoff: Bioinorganic Vision

ScienciaScripts

Imprint

Cover image: www.ingimage.com

This book is a translation from the original published under ISBN 978-620-6-72381-3.

Publisher:
Sciencia Scripts
is a trademark of
Dodo Books Indian Ocean Ltd. and OmniScriptum S.R.L publishing group

120 High Road, East Finchley, London, N2 9ED, United Kingdom
Str. Armeneasca 28/1, office 1, Chisinau MD-2012, Republic of Moldova, Europe
Printed at: see last page
ISBN: 978-620-8-24074-5

Wernicke-Korsakoff: Bioinorganic Vision

Author : Dr. Abdelhafid Mimouni

An independent researcher in bioinorganic chemistry, Dr Mimouni is an expert in macromolecular synthesis and characterisation. He obtained his PhD in Chemistry from the University of Paris XII in 1997, after a Diplôme des Études Approfondies in Bioinorganic Systems from the University of Paris XI in 1993, where he also obtained his Licence and Maîtrise in Chemistry.

Summary: Wernicke-Korsakoff Syndrome Deciphered by Bioinorganics is a scientific exploration of the biochemical mechanisms underlying this neurological pathology, often associated with vitamin B1 (thiamine) deficiency. The book examines how bioinorganics, as a discipline, provides insight into the crucial interactions between metals, enzymes and cellular structures involved in this syndrome. By delving into the roles of metal cofactors and biochemical imbalances, the text reveals the molecular processes that lead to severe symptoms such as amnesia, confusion and ataxia. Finally, it offers innovative perspectives for research and avenues for new therapeutic approaches, providing an enriched understanding of this complex disease.

Book outline :

Introduction

Presentation of the subject

Vitamin B1, or thiamine, is an essential nutrient that plays a crucial role in energy metabolism and nervous system function. Its importance is not limited to its direct biological functions; it is also influenced by various inorganic elements, such as minerals, which are necessary for its optimal metabolism. Understanding this interaction between bioinorganics and thiamine is fundamental to grasping the implications of vitamin B1 deficiency, particularly with regard to serious pathologies such as Wernicke's encephalopathy.

Importance of vitamin B1 and bioinorganics

Thiamine is essential for converting carbohydrates into energy, making it a key element in the proper functioning of cells, particularly those in the nervous system. Bioinorganics include minerals and inorganic elements involved in various biological processes. Their interaction with thiamine is vital for ensuring efficient metabolism and preventing neurological disorders.

Aims of the book

This book aims to explore in depth the role of bioinorganics in vitamin B1 metabolism and to examine the consequences of thiamine deficiency, with a particular focus on Wernicke's encephalopathy. Through a detailed analysis of biological mechanisms, case studies and practical recommendations, we aim to provide a comprehensive understanding of this crucial public health topic.

Chapter 1: Vitamin B1 (Thiamine)

Chemical structure and properties

Thiamine, or vitamin B1, is an essential water-soluble vitamin that plays a key role in energy metabolism. Its chemical structure comprises two nuclei: a pyrimidine nucleus and a thiazole nucleus, linked by a methylene chain. This configuration gives it unique properties:

Solubility: As a water-soluble vitamin, thiamine dissolves easily in water, making it easier to absorb from the gastrointestinal tract.

Sensitivity: Thiamine is sensitive to heat, oxidation and acidity, which can lead to its breakdown during cooking or food storage. Consequently, food preparation methods can influence available thiamine levels.

Food sources

Thiamine is found in a variety of foods, and a balanced diet is crucial to ensure adequate intake. The main food sources include :

Wholegrain cereals: Brown rice, wholemeal bread and rolled oats are rich in thiamine, especially when they are unrefined.

Pulses: Chickpeas, lentils and beans are excellent sources of thiamine.

Meat: Pork and offal, such as liver, contain high concentrations of thiamine.

Nuts and seeds: Almonds and sunflower seeds also provide a significant amount of thiamine.

Dairy products: Although less concentrated, milk and yoghurt contribute to the daily intake.

Physiological role in the body

Thiamine has a number of essential physiological functions, including :

Energy metabolism :

Enzyme cofactor: Thiamine is a key cofactor for several enzymes, in particular pyruvate dehydrogenase, which catalyses the conversion of pyruvate into acetyl-CoA. This process is fundamental to the Krebs cycle, where acetyl-CoA is used to produce energy in the form of ATP.

Breakdown of carbohydrates: Thiamine is involved in glycolysis, where it helps to break down glucose into energy, which is essential for cells.

Functioning of the nervous system :

Neurotransmitter synthesis: Thiamine is necessary for the synthesis of neurotransmitters such as acetylcholine, which is crucial for the transmission of nerve signals.

Neuronal protection: It plays a role in protecting neurons against oxidative damage, thereby contributing to brain health.

Regulation of carbohydrate metabolism :

Metabolic intermediate: Thiamine helps regulate glucose levels in the blood by facilitating its use by cells. This is particularly important for glucose-sensitive tissues such as the brain and muscles.

Interaction with other nutrients : Thiamine interacts with other B vitamins, such as riboflavin and niacin, to optimise overall energy metabolism.

Role in cardiovascular health :

Heart function: Thiamine is essential for the proper functioning of the heart, as it helps maintain adequate energy levels in heart cells. A deficiency can lead to heart problems such as Weternicke-Korsakoff syndrome.

Chapter 2: Metabolism of vitamin B1

Introduction

Vitamin B1, also known as thiamine, plays a crucial role in cellular energy metabolism. This chapter focuses on the different stages of thiamine metabolism, from its ingestion through the processes of digestion, absorption and conversion to its active role in the form of thiamine pyrophosphate (TPP). We will also explore the main metabolic pathways and enzymatic functions associated with vitamin B1.

Digestion and absorption processes

Ingestion and Food Sources

Vitamin B1 is mainly ingested through food sources such as wholemeal cereals, meat (particularly pork), legumes, nuts and certain vegetables. The thiamine present in food is often in the form of thiamine phosphate, which needs to be processed to be absorbed effectively.

Gastric and intestinal digestion

Digestion of thiamine begins in the stomach, where gastric acidity releases thiamine from dietary protein complexes. It is then absorbed mainly from the small intestine, in the duodenum and jejunum, by an active transport mechanism at low concentrations, and by passive diffusion at higher concentrations.

Blood Transport

Once absorbed, thiamine is transported in the bloodstream in free form or bound to plasma proteins. It is rapidly taken up by cells, particularly those in the liver, heart and brain, where it is converted into its active form.

Conversion to Thiamine Pyrophosphate (TPP)

Thiamine, or vitamin B1, is a water-soluble vitamin essential for many metabolic processes in the body. However, to carry out its biological functions, thiamine must be converted into its active coenzyme form, thiamine pyrophosphate (TPP). This conversion process takes place mainly at intracellular level and involves a series of precise enzymatic reactions.

Intracellular phosphorylation

The conversion of thiamine to TPP is a crucial process that occurs inside cells, with significant concentrations in the liver, but also in other tissues such as the brain and muscles. This process begins with thiamine entering cells via specific transporters in the cell membrane. Once inside the cell, thiamine undergoes phosphorylation, a biochemical reaction in which a phosphate group is added to the thiamine molecule.

This phosphorylation is catalysed by the enzyme thiamine pyrophosphokinase (TPK). This enzyme plays a central role in the bioactivation of thiamine, using ATP (adenosine triphosphate) as the phosphate donor. The reaction produces a molecule of TPP and AMP (adenosine monophosphate) as a by-product.

The reaction can be described as follows:

Thiamine+2ATP→Thiamine pyrophosphate (TPP)+2ADP

Control of thiamine pyrophosphokinase activity is essential for regulating the levels of TPP available in cells, which in turn influences many TPP-dependent metabolic pathways. Thiamine deficiency or abnormalities in the activity of this enzyme can lead to serious metabolic dysfunctions, such as Wernicke's encephalopathy.

Distribution of TPP

Once formed, TPP is distributed throughout various tissues in the body, where it plays a key role as a coenzyme in a number of enzymatic reactions. The distribution of TPP is regulated according to the energy and metabolic needs of the various organs.

The brain is one of the main consumers of TPP. TPP is essential for oxidative decarboxylation reactions, such as the conversion of pyruvate into acetyl-CoA, a key step in the entry of carbon into the Krebs cycle, which is essential for the production of energy in the form of ATP. The brain, being an energy-intensive organ, is highly dependent on the supply of TPP to maintain its metabolic functions.

In addition, TPP is vital for carbohydrate metabolism in muscles. Skeletal muscles, in particular, use TPP in the glycolysis and Krebs cycle reactions, enabling muscle contraction and the maintenance of muscle strength. A deficiency in TPP can therefore lead to muscular disorders such as weakness and fatigue.

Mitochondria, which are abundant in brain and muscle cells, are the main sites of action for TPP. Here, TPP also participates in the pentose phosphate pathway, another metabolic pathway crucial for the production of NADPH and the synthesis of nucleic acids, contributing to antioxidant protection and the synthesis of fatty acids and steroids.

Finally, the distribution of TPP is also modulated by transport mechanisms within cells, ensuring that this coenzyme is available in sufficient quantities for the enzymes that depend on it. Active transport systems ensure that TPP is efficiently distributed in the intracellular compartments where it is needed, such as the mitochondria and cytoplasm.

Chapter 3. Enzymatic roles and metabolic pathways

Thiamine pyrophosphate (TPP) is a crucial coenzyme in several fundamental metabolic processes. These processes are essential for energy metabolism, the synthesis of biomolecules and the maintenance of cellular homeostasis. This chapter explores in detail the molecular mechanisms by which TPP is involved in the oxidative decarboxylation of α-keto acids, the pentose phosphate pathway, and the metabolism of branched-chain amino acids.

Oxidative decarboxylation of α-keto acids

Role of TPP in Oxidative Decarboxylation

TPP plays an essential role in the oxidative decarboxylation of α-keto acids, a key process in energy production. This reaction occurs in the mitochondria and is catalysed by several enzyme complexes, the most important of which are pyruvate dehydrogenase complex (PDH), α-ketoglutarate dehydrogenase complex (α-KGDH), and branched-chain amino acid dehydrogenase complex (BCKDH).

Pyruvate dehydrogenase complex (PDH)

The pyruvate dehydrogenase (PDH) complex catalyses the conversion of pyruvate, an end product of glycolysis, into acetyl-CoA, an essential substrate for the Krebs cycle. TPP acts by facilitating the decarboxylation of pyruvate to produce a hydroxyethyl-TPP intermediate, which is then transferred to the lipoamide of the enzyme, allowing the formation of acetyl-CoA. This process releases carbon dioxide and generates NADH, which is used in the respiratory chain to produce ATP.

α-Ketoglutarate dehydrogenase complex (α-KGDH)

The α-ketoglutarate dehydrogenase complex catalyses the conversion of α-ketoglutarate to succinyl-CoA in the Krebs cycle. TPP acts in a similar way here, facilitating the decarboxylation of α-ketoglutarate. This reaction is crucial for the continuation of the Krebs cycle, allowing the production of GTP (or ATP, depending on the cell type), as well as the generation of NADH and $FADH_2$ for mitochondrial respiration.

Branched Chain Amino Acid Dehydrogenase Complex (BCKDH)

BCKDH catalyses the oxidative decarboxylation of branched-chain amino acids (leucine, isoleucine, valine), a process essential for their catabolism. TPP, as a coenzyme in this complex, facilitates the decarboxylation of the α-ketoacids derived from these amino acids, enabling them to be converted into acyl-CoA and substrates that can be used for energy production or the synthesis of other compounds.

Pentose phosphate pathway

Introduction to the Pentose Phosphate Pathway

The pentose phosphate pathway (PPP) is an alternative metabolic pathway to glycolysis, essential for the production of NADPH and ribose-5-phosphate. NADPH is crucial for fatty acid biosynthesis reactions and antioxidant defence, while ribose-5-phosphate is a precursor in nucleotide synthesis.

Role of TPP in Transketolase

Transketolase is a key PPP enzyme that requires TPP as a coenzyme. It catalyses the transfer of two carbon units from xylulose-5-phosphate to ribose-5-phosphate to form fructose-6-phosphate and glyceraldehyde-3-phosphate. This mechanism involves intermediate stabilisation by TPP, allowing the transfer of an aldehyde group and facilitating the continuity of the metabolic pathway.

Importance of NADPH and Ribose-5-Phosphate

The NADPH produced by the pentose phosphate pathway is vital for maintaining the cell's redox balance and for biosynthetic reactions, such as the synthesis of fatty acids and cholesterol. Ribose-5-phosphate, meanwhile, is essential for nucleotide biosynthesis and, consequently, for DNA and RNA synthesis, making TPP essential for cell proliferation and survival.

Metabolism of branched-chain amino acids

Catabolism of branched-chain amino acids

The branched-chain amino acids (BCAAs) - leucine, isoleucine and valine - are essential for protein synthesis and energy production. Their catabolism begins with transamination, followed by oxidative decarboxylation catalysed by the BCKDH complex, in which TPP plays a central role.

Conversion into usable products

TPP facilitates the conversion of α-keto acids from AACR to acyl-CoA, metabolic intermediates that can enter the Krebs cycle or serve as precursors for other anabolic pathways. For example, isoleucine can be converted into succinyl-CoA and acetyl-CoA, while leucine is converted into acetyl-CoA and HMG-CoA, the latter being a precursor for the biosynthesis of ketone bodies and cholesterol.

Regulation and Clinical Importance

AACR metabolism is tightly regulated and its dysfunction can lead to pathologies such as maple syrup urine disease (MSUD), in which BCKDH is deficient. The role of TPP in this process is therefore crucial not only for energy metabolism, but also for regulating the concentration of amino acids in the blood.

Conclusion

The enzymatic roles of TPP in metabolic mechanisms are diverse and essential for energy production, biosynthesis and the regulation of cellular processes. An in-depth understanding of these mechanisms is crucial for developing targeted therapeutic strategies for diseases linked to thiamine deficiency or specific metabolic dysfunctions.

Chapter 4: Thiamine deficiency

Thiamine deficiency, also known as beriberi when severe, is a public health problem in many parts of the world, particularly in developing countries. This deficiency can have multiple causes and varied clinical manifestations, and particularly affects certain population groups. This chapter explores in depth the causes, symptoms and consequences of thiamine deficiency, as well as the groups at risk.

Causes of Thiamine Deficiency

Thiamine deficiency can result from a number of factors, mainly nutritional, pathological and other special conditions.

Nutritional causes

Diets low in thiamine are the most direct cause of deficiency. Refined foods, such as polished white rice, white bread and other cereals that have had their outer husks removed, contain little thiamine. In certain regions of South-East Asia, where white rice is the staple diet, thiamine deficiency is common. In addition, chronic alcoholism interferes with the absorption of thiamine and its conversion to TPP, increasing the risk of deficiency.

Pathological causes

Certain medical conditions increase the risk of thiamine deficiency. Diseases affecting the gastrointestinal tract, such as coeliac disease, Crohn's disease or other malabsorption disorders, can lead to reduced thiamine absorption. In addition, patients suffering from chronic renal failure undergoing dialysis are at risk, as dialysis can lead to a loss of thiamine.

Special Conditions

Thiamine requirements increase under certain physiological conditions, such as pregnancy, breast-feeding and periods of rapid growth in children. People undergoing long-term diuretic treatment or suffering from eating disorders, such as anorexia or bulimia, are also at risk of developing a thiamine deficiency.

Symptoms and Consequences of Thiamine Deficiency :

Thiamine deficiency can manifest itself through a variety of symptoms, depending on the severity of the deficiency and the organ systems affected.

Neurological symptoms (Dry beriberi): Dry beriberi primarily affects the peripheral nervous system. Patients may experience peripheral neuropathy with symptoms such as numbness, tingling, muscle weakness and impaired coordination. At an advanced stage, this can progress to Korsakoff's syndrome, characterised by memory impairment, mental confusion and hallucinations.

Cardiovascular symptoms (Wet beriberi) :

Wet beriberi is characterised by damage to the cardiovascular system. It can lead to congestive heart failure, peripheral oedema, tachycardia and low blood pressure. In severe cases, heart failure can be fatal if not treated promptly with thiamine supplements.

Wernicke's encephalopathy :

This severe form of thiamine deficiency is a medical emergency. It manifests itself as a classic clinical triad: ataxia (impaired coordination of movement), ophthalmoplegia (paralysis of the eye muscles) and mental confusion. If not treated immediately with intravenous thiamine, it can progress to Korsakoff's syndrome.

Other consequences

Prolonged thiamine deficiency can also affect digestion, energy production and carbohydrate metabolism, leading to chronic fatigue, loss of appetite and weight loss.

Risk Groups

Certain population groups are more likely to develop a thiamine deficiency, notably because of their dietary habits, medical conditions or other risk factors.

Chronic Alcoholics:

People suffering from chronic alcoholism are particularly at risk because of alcohol's interference with thiamine absorption and metabolism. Alcohol also causes inflammation of the stomach and intestines, which can reduce the absorption of nutrients.

People suffering from malnutrition :

People living in conditions of severe malnutrition, including those suffering from famine or eating disorders, or living in regions where poverty is endemic, are at risk of thiamine deficiency.

Pregnant and Breastfeeding Women:

Thiamine requirements increase during pregnancy and breastfeeding. If these needs are not met by the diet, there is an increased risk of developing thiamine deficiency, which can also affect the foetus or infant.

Dialysis patients :

Patients on chronic kidney dialysis can lose thiamine during the dialysis process, putting them at risk of deficiency.

Seniors :

The elderly are also at risk of thiamine deficiency due to reduced intestinal absorption, an often insufficient diet, and a greater prevalence of chronic diseases affecting nutrient absorption.

Chapter 5: Wernicke's encephalopathy

Wernicke's encephalopathy is a serious and potentially reversible neurological condition caused by a deficiency of thiamine (vitamin B1). It often manifests as a clinical triad of mental confusion, ataxia and ophthalmoplegia (paralysis of eye movements). Wernicke's encephalopathy is frequently associated with chronic alcoholism, which leads to malabsorption and deficient metabolism of thiamine, but it can also occur in other clinical contexts, such as eating disorders, gastrointestinal diseases, bariatric surgery and hypermetabolic conditions.

Etiology: The main cause of Wernicke's encephalopathy is thiamine deficiency. Thiamine is an essential cofactor in several metabolic pathways, including the oxidative decarboxylation of α-keto acids and the pentose phosphate pathway. In the absence of thiamine, these pathways are disrupted, resulting in decreased ATP production and lactate accumulation, leading to neuronal dysfunction and cellular damage.

The causes of thiamine deficiency can be divided into several categories:

- **Nutritional**: insufficient thiamine intake, often due to excessive alcohol consumption, which reduces intestinal absorption of the vitamin.
- **Pathological**: Diseases that interfere with intestinal absorption or increase the metabolic demand for thiamine, such as chronic infections, cancer or disorders of the gastrointestinal system.
- **Iatrogenic**: Surgery or medical treatments that affect the absorption or metabolism of thiamine.

Physiopathological mechanisms

Wernicke's encephalopathy results mainly from a thiamine deficiency which disrupts the functions of thiamine-dependent enzymes. These enzymes, such as **pyruvate dehydrogenase** and **transketolase**, are essential for energy production by neuronal cells.

- **ATP deficiency**: A lack of thiamine alters the function of pyruvate dehydrogenase, a crucial enzyme in the conversion of pyruvate to acetyl-CoA in the Krebs cycle. This disruption leads to a reduction in the production of ATP, which is necessary for neuronal function.
- **Lactate accumulation**: Because pyruvate dehydrogenase is unable to metabolise pyruvate, it is redirected to the anaerobic pathway, where it is converted to lactate. The accumulation of

lactate in the brain causes local acidosis, leading to neuronal dysfunction.

- **Impaired mitochondrial function**: Thiamine is also involved in mitochondrial function. Thiamine deficiency can lead to increased oxidative stress and cell death, exacerbating brain damage.

Cerebral lesions are mainly observed in regions rich in glial cells, such as the mammillary bodies, the thalamus and the brain stem, which are particularly vulnerable to energy disturbances.

Clinical symptoms and diagnosis

The clinical symptoms of Wernicke's encephalopathy are classically described by the triad:

- **Mental confusion**: Altered mental state is the most common symptom. It can vary from slight disorientation to a state of coma.
- **Ataxia**: Patients often present with an unsteady gait due to cerebellar dysfunction. Ataxia may be associated with dysarthria and impaired motor coordination.
- **Ophthalmoplegia**: Ocular paralysis, characterised by nystagmus and paresis of the eye muscles, is a distinctive sign. This

ophthalmoplegia is due to lesions of the oculomotor nuclei in the brain stem.

The diagnosis of Wernicke's encephalopathy is primarily clinical, based on recognition of the symptomatic triad. However, magnetic resonance imaging (MRI) can reveal characteristic lesions in affected brain regions, such as the mammillary bodies and thalamus. Prompt intravenous administration of thiamine is the treatment of choice and can reverse symptoms, especially if administered at an early stage of the disease.

Chapter 6: Link between Bioinorganics and Wernicke's encephalopathy

Impact of Mineral Deficiencies on Thiamine Metabolism

The metabolism of thiamine, an essential water-soluble vitamin, is closely linked to the state of minerals in the body. Minerals play a crucial role in the biochemical processes that enable thiamine to be converted into its active form, thiamine pyrophosphate (TPP), and to perform its vital enzymatic functions. Several inorganic elements, notably magnesium, zinc and iron, are involved in these processes.

- **Magnesium**: Magnesium is an essential cofactor in the phosphorylation of thiamine to TPP by the enzyme thiamine pyrophosphokinase. Magnesium deficiency can interfere with this conversion, reducing the availability of active thiamine for enzymatic reactions. This can exacerbate the effects of thiamine deficiency and contribute to the development of Wernicke's encephalopathy.
- **Zinc**: Zinc is necessary for the function of several thiamine-dependent enzymes. Zinc deficiency can lead to a reduction in the effectiveness of these enzymes, aggravating the symptoms associated with thiamine deficiency. In addition, zinc plays a role in maintaining neuronal integrity and protecting against oxidative stress, aspects which are critical in preventing the brain damage associated with Wernicke's encephalopathy.

- **Iron**: Iron is involved in cellular respiration and ATP production, processes in which thiamine also plays a crucial role. An iron deficiency can compromise energy production in cells, exacerbating the effects of thiamine deficiency and increasing the risk of developing Wernicke's encephalopathy.

The role of inorganic elements in prevention and treatment

Inorganic elements not only affect thiamine metabolism, they also play a crucial role in the prevention and treatment of Wernicke's encephalopathy.

- **Magnesium**: As an essential cofactor in the conversion of thiamine, an adequate intake of magnesium is crucial in preventing deficiencies in TPP. In the treatment of Wernicke's encephalopathy, the administration of magnesium as a supplement to thiamine can improve the effectiveness of therapy and speed up recovery. Clinical studies have shown that patients receiving both thiamine and magnesium showed a more rapid improvement in neurological symptoms than those receiving thiamine alone.
- **Zinc and antioxidants**: Because of its role in protecting against oxidative stress, zinc can help limit neuronal damage in thiamine deficiency. Incorporating zinc into treatment protocols can help

boost the brain's antioxidant defences, thereby reducing the severity of symptoms. Zinc supplements are often recommended in cases of malnutrition or chronic alcoholism, major risk factors for Wernicke's encephalopathy.

- **Iron and energy metabolism**: By preventing anaemia and supporting ATP production, sufficient iron intake can mitigate the effects of thiamine deficiency. However, iron supplementation must be carefully monitored, as excess iron can generate oxidative stress, potentially harmful to vulnerable neurons.

Case Studies and Recent Research

Recent research continues to elucidate the complex link between inorganic elements and Wernicke's encephalopathy. Several case studies have demonstrated the importance of minerals in the management of this condition.

- **Clinical studies on magnesium**: Recent studies have shown that the administration of magnesium in parallel with thiamine in the treatment of Wernicke's encephalopathy can lead to a better recovery of cognitive functions. These studies suggest that magnesium potentiates the efficacy of thiamine, probably by improving its conversion to TPP and supporting thiamine-dependent metabolic pathways.

- **Protective role of Zinc**: Research has shown that patients with chronic gastrointestinal diseases, who are often zinc deficient, have an increased risk of developing Wernicke's encephalopathy. Zinc supplementation in these populations has been shown to be beneficial in reducing the incidence of this condition.
- **Multimodal interventions**: Treatments combining thiamine, magnesium, zinc and iron, combined with overall nutritional management, have been proposed as an effective strategy for preventing Wernicke's encephalopathy in at-risk populations, such as patients suffering from eating disorders or chronic alcoholics.

These studies underline the importance of a bioinorganic approach to the management of Wernicke's encephalopathy, where mineral balance plays as crucial a role as thiamine supplementation itself.

Chapter 7: Prevention and treatment

Nutritional Strategies to Prevent Thiamine Deficiency

Preventing thiamine deficiency depends largely on a balanced diet that ensures an adequate intake of this essential vitamin. Nutritional strategies must be tailored to the specific needs of individuals, taking into account risk factors such as age, sex, general state of health and eating habits.

- **Balanced diet**: A diet rich in natural sources of thiamine is essential. Foods such as wholemeal cereals, legumes, pork, nuts and seeds are excellent sources of thiamine. Regular consumption of these foods can go a long way towards preventing deficiency.
- **Specific diets**: For at-risk groups, such as the elderly, alcoholics or those suffering from chronic illnesses, specific diets may be necessary. For example, people suffering from intestinal malabsorption could benefit from a thiamine-enriched diet and regular monitoring of their vitamin levels.
- **Food fortification**: In certain populations at high risk of thiamine deficiency, such as those living in regions with a poorly varied diet, fortifying staple foods with thiamine has proved to be an effective strategy. Examples of such fortified foods include bread, cereals and dairy products.

Importance of Thiamine and Mineral Supplementation

Thiamine supplementation is crucial for preventing and treating thiamine deficiency, particularly in people with an increased risk or early signs of deficiency. Minerals such as magnesium, zinc and iron also play an important complementary role.

- **Thiamine supplementation**: In at-risk individuals or those with a diagnosed deficiency, thiamine supplementation in the form of tablets or injections may be necessary. Doses may vary according to the severity of the deficiency, but high doses are often required to treat severe neurological symptoms, such as those seen in Wernicke's encephalopathy.
- **Magnesium**: Magnesium supplementation is often required alongside thiamine to optimise its metabolism in TPP. Magnesium is essential for the activity of thiamine pyrophosphokinase, the enzyme responsible for thiamine phosphorylation. Therefore, in the event of magnesium deficiency, thiamine supplementation alone may not be sufficient to correct the symptoms.
- **Zinc and iron**: Zinc and iron also play essential roles in thiamine metabolism and protection against cell damage. Zinc, as an antioxidant, helps protect neurons against oxidative stress, while

iron is necessary for energy metabolism, a process in which thiamine is directly involved.

Therapeutic Approaches for Wernicke's Encephalopathy

Wernicke's encephalopathy is a medical emergency requiring immediate treatment to prevent irreversible brain damage. Treatment focuses on rapid correction of thiamine deficiency and management of associated complications.

- **Urgent administration of thiamine**: Treatment of Wernicke's encephalopathy begins with the rapid administration of thiamine, usually intravenously to ensure immediate and effective absorption. High doses of thiamine (500 mg or more per day) are often given to overcome the blood-brain barrier and treat acute neurological symptoms.
- **Magnesium supplementation**: As a supplement to thiamine, intravenous or oral administration of magnesium is recommended. This not only improves the effectiveness of thiamine, but also helps prevent muscle spasms, convulsions and other neurological complications associated with magnesium deficiency.
- **Multimodal treatment**: In addition to thiamine and mineral supplementation, treatment of Wernicke's encephalopathy may

include overall nutritional management, correction of electrolyte imbalances, and management of associated complications, such as liver failure in alcoholics. Cognitive and neurological re-education may also be necessary for patients who have suffered significant brain damage.

- **Long-term follow-up**: After the acute phase, long-term follow-up is essential to prevent relapses and manage the potential sequelae of Wernicke's encephalopathy. This includes ongoing supplementation with thiamine and magnesium, as well as nutritional adjustments to maintain adequate levels of these nutrients.

Conclusion

Summary of Key Points

Thiamine, or vitamin B1, plays a fundamental role in energy metabolism and neurological function. Previous chapters have detailed the mechanisms by which thiamine is absorbed, converted to its active form, thiamine pyrophosphate (TPP), and integrated into crucial metabolic pathways such as the oxidative decarboxylation of α-keto acids and the pentose phosphate pathway. The impact of thiamine deficiency, particularly Wernicke's encephalopathy, has also been explored, highlighting the severe neurological consequences that can result from prolonged deficiency.

We also examined preventive and therapeutic strategies for maintaining optimal thiamine levels, emphasising the importance of appropriate supplementation and a thiamine-rich diet. The integration of minerals such as magnesium, zinc and iron was identified as a key factor in supporting the efficacy of thiamine in its biochemical roles.

The importance of an integrated approach to nutrition

Preventing thiamine deficiency does not rely solely on one-off supplementation or on correcting existing deficiencies. An integrated approach to nutrition, which considers the interactions between vitamins and minerals, is essential to ensure optimal health. For example, thiamine metabolism is intrinsically linked to the availability of magnesium, which is required for the conversion of thiamine to TPP. Similarly, the availability of zinc and iron can influence the efficiency of the enzymatic processes in which thiamine is involved.

This integrated approach also involves ongoing assessment of dietary intakes, particularly for groups at risk of deficiency, such as the elderly, chronic alcoholics, and patients suffering from malabsorption or chronic illnesses. In addition, nutritional education needs to be stepped up to raise awareness among the public and healthcare professionals of the importance of maintaining an adequate vitamin and mineral balance.

Future Research Perspectives

Although the basics of thiamine metabolism are well established, there are several areas for future research. Firstly, more in-depth studies are needed to understand the specific interactions between thiamine and other nutrients, particularly in pathological contexts where metabolism is disturbed.

Secondly, the development of new supplementation strategies could improve the effectiveness of the prevention and treatment of thiamine deficiency. For example, exploring bioavailable forms of thiamine and developing supplementation formulas that incorporate essential minerals could offer better options for patients at risk.

Finally, research into the long-term implications of subclinical thiamine deficiency could reveal hitherto unexplored links with other neurodegenerative or metabolic diseases. This could pave the way for new preventive approaches in the field of public health.

Lexicon

- **Thiamine (Vitamin B1)** : Water-soluble vitamin essential for energy metabolism and the proper functioning of the nervous system. It must be converted into Thiamine Pyrophosphate (TPP) to become active in enzymatic reactions.
- **Bioinorganics**: Area of chemistry that studies the interactions between inorganic elements and biological systems.
- **Wernicke's encephalopathy**: Serious neurological disorder caused by thiamine deficiency, often associated with alcoholism, characterised by a triad of ataxia, ophthalmoplegia and mental confusion.
- **Cofactor**: Non-protein substance or molecule required for the activity of an enzyme.
- **Metabolism**: All the chemical reactions that take place in cells to maintain life.
- **Hydrosoluble**: Refers to a substance that dissolves in water.
- **Pyruvate**: End product of glycolysis, essential for energy production via its conversion into Acetyl-CoA.
- **Acetyl-CoA**: Key molecule in energy metabolism, serving as a substrate in the Krebs cycle.
- **Krebs cycle**: Series of biochemical reactions in the mitochondria, essential for the production of energy in the form of ATP.

- **Phosphorylation**: Process of adding a phosphate group to a molecule, often to activate it. This includes the conversion of thiamine to TPP.
- **Oxidative decarboxylation**: chemical reaction in which a molecule loses a carboxyl group in the form of CO_2, catalysed by enzymes requiring TPP as a cofactor.
- **Pentose phosphate pathway**: A metabolic pathway parallel to glycolysis which generates NADPH and pentoses for nucleic acid biosynthesis.
- **Branched Chain Amino Acids (BCAAs)**: Essential amino acids such as leucine, isoleucine and valine, involved in protein synthesis and energy metabolism.
- **α-Ketonic acids**: Important metabolic intermediates in oxidative decarboxylation, such as pyruvate and α-ketoglutarate.
- **Biosynthesis**: Process by which cells produce complex molecules from simple precursors, as in the pentose phosphate pathway.
- **Lipoamide**: Prosthetic group involved in oxidative decarboxylation reactions, such as those catalysed by pyruvate dehydrogenase.
- **NADPH (Nicotinamide Adenine Dinucleotide Phosphate)** : Coenzyme used in antioxidant biosynthesis and regeneration reactions, produced mainly by the pentose phosphate pathway.

- **Pentoses**: Five-carbon sugars produced in the pentose phosphate pathway, used in the biosynthesis of nucleic acids.
- **Transketolase**: Enzyme of the pentose phosphate pathway which catalyses the transfer of carbon groups between phosphate sugars, requiring TPP as a coenzyme.
- **ATP (Adenosine Triphosphate)**: The main energy-carrying molecule in cells, involved in the phosphorylation of thiamine to form TPP.
- **ADP (Adenosine Diphosphate)**: Molecule produced when ATP is dephosphorylated, releasing energy for cellular processes.
- **AMP (Adenosine Monophosphate)** : By-product of the conversion of ATP to ADP, playing a role in the regulation of cellular energy.
- **Pyruvate dehydrogenase complex**: set of enzymes which catalyse the conversion of pyruvate into acetyl-CoA, a key reaction in energy metabolism, with TPP as the essential cofactor.
- **Thiamine deficiency**: Condition resulting from insufficient thiamine intake, leading to disorders such as beriberi or Wernicke's encephalopathy.
- **Glycolysis**: Metabolic pathway which converts glucose into pyruvate, producing energy in the form of ATP. Pyruvate is then converted into acetyl-CoA to enter the Krebs cycle.

- **Mitochondria**: Cellular organelles responsible for producing energy through cellular respiration, where TPP facilitates the reactions of the Krebs cycle.
- **Thiamine Pyrophosphokinase (TPK)**: Enzyme which catalyses the phosphorylation of thiamine to convert it into TPP, essential for the metabolism of thiamine in cells.
- **Ataxia**: Disorder in the coordination of muscular movements, often observed in Wernicke's encephalopathy.
- **Beriberi**: Disease resulting from severe thiamine deficiency, with dry (neurological) or wet (cardiovascular) forms.
- **Dry beriberi** : A form of beriberi primarily affecting the peripheral nervous system, causing neurological symptoms such as peripheral neuropathy.
- **Humid beriberi**: A form of beriberi mainly affecting the cardiovascular system, leading to heart failure and oedema.
- **Celiac disease**: An autoimmune disease in which ingestion of gluten damages the small intestine, leading to malabsorption of nutrients, including thiamine.
- **Crohn's disease**: Chronic inflammatory disease of the intestine which can affect the absorption of nutrients and increase the risk of thiamine deficiency.

- **Peripheral neuropathy**: Damage to the peripheral nerves causing numbness, tingling and muscle weakness, often linked to thiamine deficiency.
- **Korsakoff's syndrome**: Serious neurological disorder, often a complication of untreated Wernicke's encephalopathy, characterised by memory impairment and hallucinations.
- **Tachycardia**: Abnormally rapid heart rate, which may be a symptom of wet beriberi.
- **Acidosis**: A condition characterised by an increase in acidity in the tissues and blood, often due to the accumulation of lactate.
- **Ophthalmoplegia**: Paralysis of the eye muscles, often associated with eye movement disorders.
- **Magnesium**: Essential mineral involved in many biochemical processes, including the conversion of thiamine to TPP.
- **Oxidative stress**: Cellular damage caused by free radicals, often linked to chronic diseases.

References :

- Bender, D. A. (2014). *Nutritional Biochemistry*. Academic Press.
- Butterworth, R. F. (2003). Thiamine deficiency-related brain dysfunction in chronic liver failure. *Metabolic Brain Disease, 18*(3), 205-212. https://doi.org/10.1023/A:1025547714533
- Butterworth, R. F. (2005). Thiamine deficiency and brain disorders. In *Handbook of Clinical Neurology* (Vol. 83, pp. 828-835). Elsevier. https://doi.org/10.1016/S0072-9752(06)83068-5
- Caine, D., Halliday, G. M., Kril, J. J., & Harper, C. G. (1997). Operational criteria for the classification of chronic alcoholics: Identification of Wernicke's encephalopathy. *Journal of Neurology, Neurosurgery & Psychiatry, 62*(1), 51-60. https://doi.org/10.1136/jnnp.62.1.51
- Combs, G. F. Jr. 2012. *The Vitamins: Fundamental Aspects in Nutrition and Health* (4th ed.). Academic Press.
- Duffy, M. (2019). *Vitamins and Minerals: A Practical Guide.* Wiley.
- Fattal-Valevski, A. (2011). Thiamine (vitamin B1). *Journal of Neuro-Ophthalmology, 31*(4), 424-428. https://doi.org/10.1097/WNO.0b013e318238b60e
- Frank, R. A., Leeper, F. J., & Luisi, B. F. (2007). Structure, mechanism and catalytic duality of thiamine-dependent enzymes.

Cellular and Molecular Life Sciences, 64(7-8), 892-905. https://doi.org/10.1007/s00018-007-6529-9

- Gibson, G. E., & Zhang, H. (2002). Interactions of oxidative stress with thiamine homeostasis promote neurodegeneration. *Neurochemistry International, 40*(6), 493-504. https://doi.org/10.1016/S0197-0186(01)00132-4
- Gubler, C. J. (1991). Vitamin B1 (Thiamin). In L. J. Machlin (Ed.), *Handbook of Vitamins* (2nd ed., pp. 233-284). Marcel Dekker.
- Haller, J. (2020). The Role of Thiamine in Metabolism and Health. *Journal of Nutritional Science, 9*(1), 1-12. https://doi.org/10.1017/jns.2020.1
- Harper, C. G., & Matsumoto, I. (2005). Ethanol and brain damage. *Current Opinion in Pharmacology, 5*(1), 73-78. https://doi.org/10.1016/j.coph.2004.09.007
- Harper, C., & Butterworth, R. (1998). Nutritional and metabolic encephalopathies. *Seminars in Neurology, 18*(2), 389-391. https://doi.org/10.1055/s-2008-1040947
- Lonsdale, D. (2006). *Thiamine Deficiency Disease, Dysautonomia, and High Calorie Malnutrition*. CRC Press.
- Lonsdale, D. (2006). Thiamine, the Brain and Metabolism. *Journal of Orthomolecular Medicine, 21*(1), 45-50.

- Lonsdale, D. (2006). A review of the biochemistry, metabolism, and clinical benefits of thiamin and its derivatives. *Evidence-Based Complementary and Alternative Medicine, 3*(1), 49-59. https://doi.org/10.1093/ecam/nek009
- Luong, K. V., & Nguyen, L. T. (2013). The impact of thiamine treatment in Alzheimer's disease. *American Journal of Alzheimer's disease & Other Dementias, 28*(4), 390-392. https://doi.org/10.1177/1533317513488914
- Manzetti, S., Zhang, J., & van der Spoel, D. (2014). Thiamin function, metabolism, uptake, and transport. *Biochemistry, 53*(5), 821-835. https://doi.org/10.1021/bi401618y
- Martin, P. R., Singleton, C. K., & Hiller-Sturmhöfel, S. (2003). The role of thiamine deficiency in alcoholic brain disease. *Alcohol Research & Health, 27*(2), 134-142.
- Oudman, E., Wijnia, J. W., Oey, M. J., Van Dam, M., Biter, L. U., & Postma, A. (2021). Preventing Wernicke encephalopathy in patients at risk: What is the best protocol? *Frontiers in Neurology, 12*, 634736. https://doi.org/10.3389/fneur.2021.634736
- Rindi, G. (1996). Thiamine absorption and metabolism. *Biochemical and Biophysical Research Communications, 224*(2), 718-721. https://doi.org/10.1006/bbrc.1996.1075

- Sechi, G., & Serra, A. (2007). Wernicke's encephalopathy: New clinical settings and recent advances in diagnosis and management. *The Lancet Neurology, 6*(5), 442-455. https://doi.org/10.1016/S1474-4422(07)70104-7
- Sharman, M. J. (2018). *Bioinorganic Chemistry in Health and Disease*. Springer.
- Singleton, C. K., & Martin, P. R. (2001). Molecular mechanisms of thiamine utilization. *Current Molecular Medicine, 1*(2), 197-207. https://doi.org/10.2174/1566524013363161
- Thompson, K., & Proctor, M. (2020). Thiamine deficiency. *StatPearls [Internet].* https://www.ncbi.nlm.nih.gov/books/NBK493158/
- Victor, M., Adams, R. D., & Collins, G. H. (1989). *The Wernicke-Korsakoff Syndrome and Related Neurologic Disorders Due to Alcoholism and Malnutrition*. F.A. Davis.

FSC
www.fsc.org
MIX
Papier aus verantwortungsvollen Quellen
Paper from responsible sources
FSC® C105338

Printed by Books on Demand GmbH, Norderstedt / Germany